U0166505

双语版 · 全4册

BUTTERFLY

亲亲动物

如果蝴蝶来我家

[英]弗朗西斯 · 罗杰斯 [英]本 · 克里斯戴尔 著绘 范晓星 译 朱朝东 丁亮 审校

中信出版集团 | 北京

图书在版编目（CIP）数据

如果蝴蝶来我家 : 汉文、英文 / (英) 弗朗西斯·罗杰斯, (英) 本克里斯戴尔著绘 ; 范晓星译. -- 北京 : 中信出版社, 2023.3

（DK亲亲动物 : 双语版 : 全4册）

ISBN 978-7-5217-5239-7

Ⅰ. ①如… Ⅱ. ①弗… ②本… ③范… Ⅲ. ①蝶一少儿读物一汉、英 Ⅳ. ①Q964-49

中国国家版本馆CIP数据核字（2023）第021864号

Original Title: How can I help Roxy the butterfly?

如果蝴蝶来我家

（DK 亲亲动物双语版 全 4 册）

著 绘：[英] 弗朗西斯 · 罗杰斯 [英] 本 · 克里斯戴尔

译 者：范晓星

出版发行：中信出版集团股份有限公司

（北京市朝阳区东三环北路 27 号嘉铭中心 邮编 100020）

承 印 者：北京顶佳世纪印刷有限公司

开 本：889mm × 1194mm 1/20　印 张：2　字 数：115 千字

版 次：2023 年 3 月第 1 版　印 次：2023 年 3 月第 1 次印刷

京权图字：01-2022-4478　审 图 号：GS 京（2022）1525号（本书插图系原文插图）

书 号：ISBN 978-7-5217-5239-7

定 价：156.00 元（全 4 册）

出 品 中信儿童书店

图书策划 红披风

策划编辑 陈瑜

责任编辑 袁慧

营销编辑 易晓倩 李鑫橦 高铭霞

装帧设计 哈_哈

服务热线：400-600-8099

投稿邮箱：author@citicpub.com

致所有好奇的孩子！

For the curious

www.dk.com

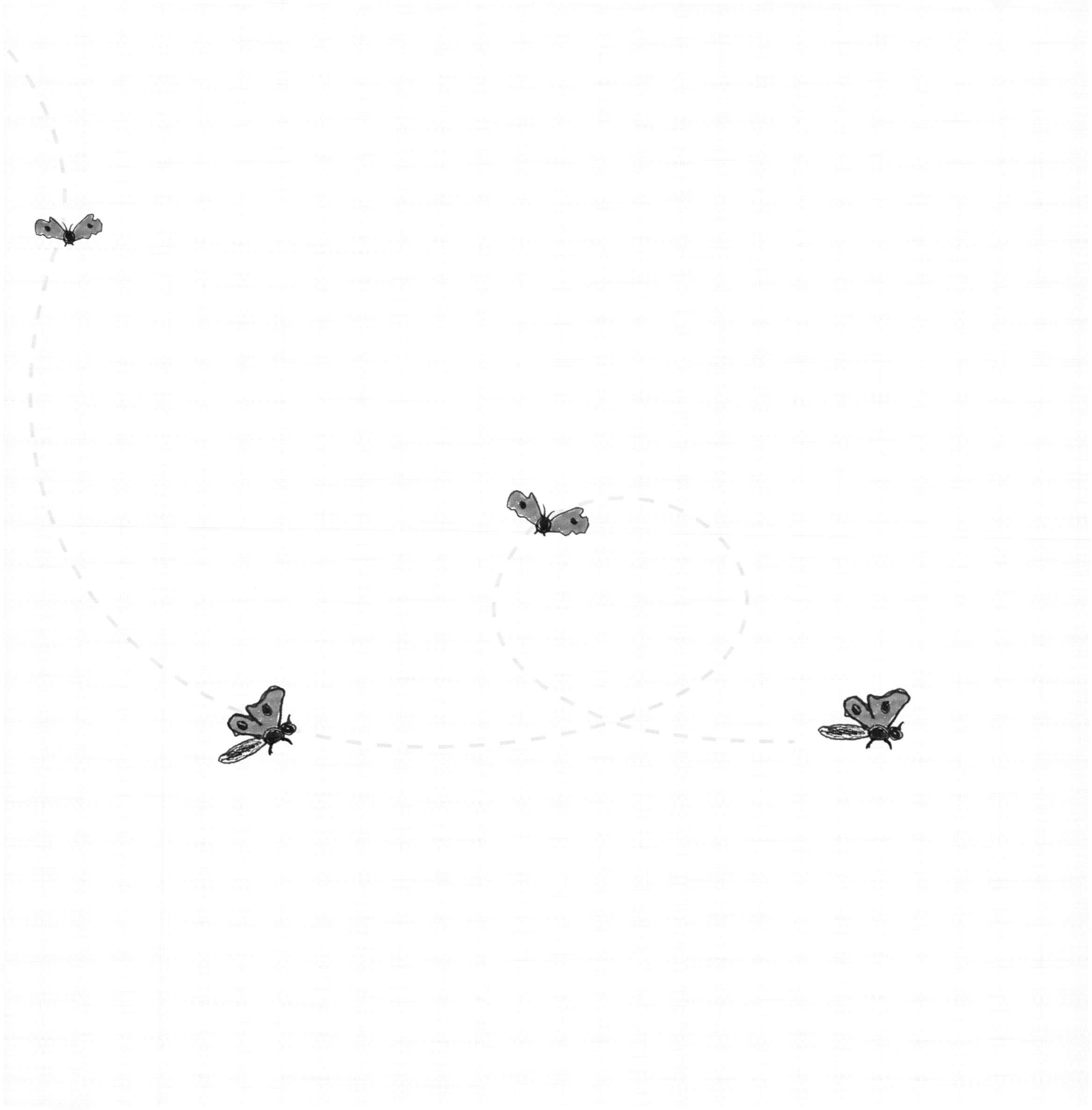

你好，我叫罗克西。
我是一只蝴蝶。
我想请你帮帮忙，可以吗？

Hello, my name is Roxy.
I am a butterfly and I need your help.

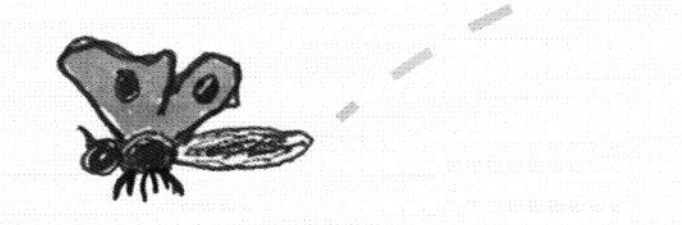

我小时候是一条毛毛虫。

I start my life as a caterpillar.

这是我的蛹，它藏在一个安全的地方，
我在里面睡大觉。

I go to sleep in my
chrysalis in a safe place.

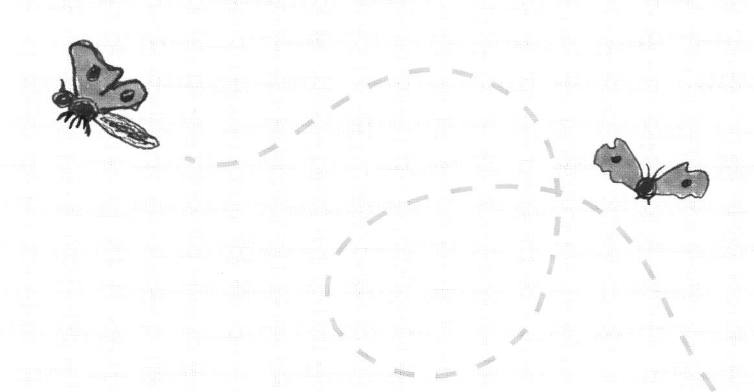

后来，我慢慢开始变化了。

Then I begin to change.

见证奇迹的时刻！
ABRACADABRA !

啊！打个哈欠！
YAWN!
伸个懒腰！
STRETCH!

我睡醒了，变成了一只蝴蝶！

When I wake up, I have
turned into a butterfly.

我想到你家花园去寻找花朵，
我喜欢吃花蜜。

I like to feed on nectar found
on the flowers in your garden.

花蜜
Nectar

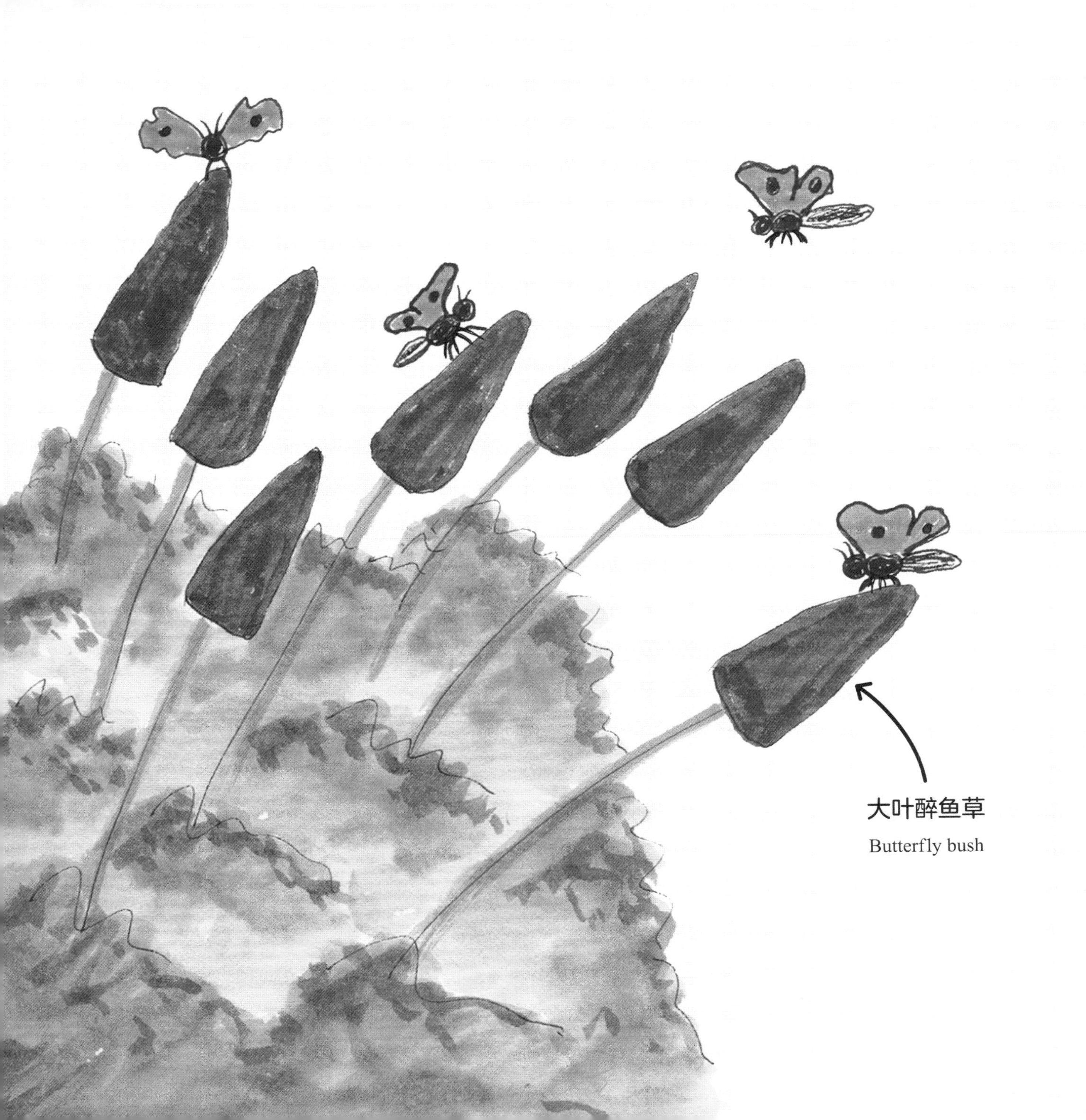
大叶醉鱼草
Butterfly bush

请你为我多种一些花吧。
我最喜欢的植物叫大叶醉鱼草。

Please plant lots of flowers for me.
My favourite plant is called Butterfly bush.

我是用脚来品尝味道的。

I can taste through my feet.

嗯！
MMM!
真香!
YUM!

我渴了，我最喜欢的饮料是果汁。

When I get thirsty, I like
to drink juice from fruit.

请在你家花园里放一些熟透的水果，
我可喜欢苹果、橙子和香蕉了。

Please put some old fruit out for
me in your garden. I like apples,
oranges, and bananas.

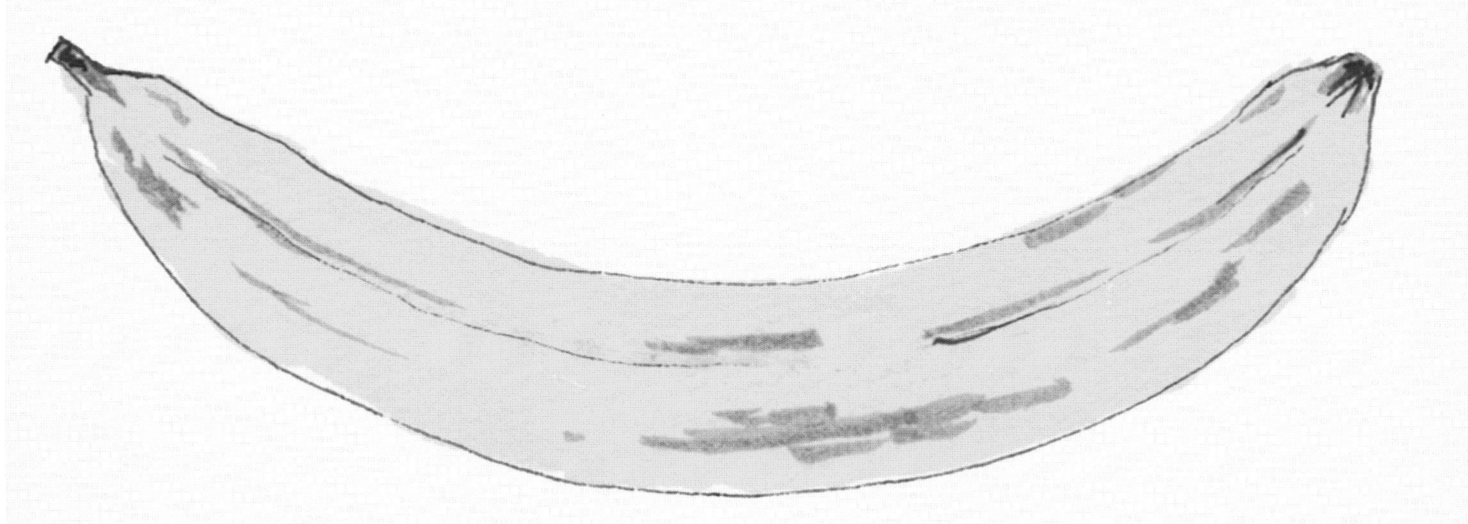

我的“舌头”有点像吸管。

My tongue is a bit like a straw.

好累啊，到哪里休息呢？

I get tired and need
somewhere to rest.

噜呼噜呼

请用木头为我搭建一座虫虫旅馆，
好不好？

Please build me a bug hotel
out of logs and wood.

谢谢你为了保护我们做的一切。

Thank you for all your help.

我们为什么要保护蝴蝶？

Why do we need to protect butterflies?

像罗克西这样的蝴蝶需要我们的保护。它们生活在世界各地。有些品种的蝴蝶已经在地球上存在很多年了——至少 5000 万年！然而，很多蝴蝶面临危险。

Butterflies, like Roxy, need to be protected. They can be found all around the world. Some types of butterflies have been around for a long time – at least 50 million years! However, many butterflies are endangered.

科学家担心这些美丽的、长着翅膀的小精灵会逐渐消失，所以我们要尽力保护它们，否则就来不及了！

Scientists are worried because some of these wonderful, winged creatures appear to be dying out. That is why we need to do our bit to help them before it is too late!

你能帮帮我们吗？
Please can you help us?

传粉

Pollination

你家花园里的某些植物会产生花蜜，花蜜是很甜的、含糖的液体。很多昆虫，包括罗克西，都喜欢吃花蜜。当蝴蝶落在你家花园的花上时，它们用长长的、吸管一样的“舌头”（口器）去吸吮花蜜。同时，花粉也就沾在了它们的身上。

Some of the plants in your garden produce nectar which is a sweet, sugary liquid that many insects, including Roxy, like to eat. When butterflies land on flowers in your garden, they use their long, straw-like tongues (proboscises) to suck up the nectar. They also get pollen stuck to their bodies.

花粉就这样被带到其他花上，这样可以使植物长出种子和果实。这个过程叫作传粉。像罗克西这样的蝴蝶就是非常重要的传粉者。在你家的花园里，你有没有见过它们？

This pollen is then carried to other flowers, which causes new seeds or fruit to grow. This process is called pollination. Butterflies, like Roxy, are important pollinators. Can you spot any in your garden?

大多数蛾在白天休息，夜晚出来。

Most moths rest in the day and come out at night.

蛾

Moth

蛾有短的、羽毛状的触角，蝴蝶有长的、细细的触角，触角的顶端有个凸起。

Moths have short, feathery antennae, whereas butterflies have long and thin antennae with a club at the end.

蝴蝶和蛾都有长长的、管状的“舌头”（口器），可以吸吮花蜜。

Both have a long, tube-like tongue (proboscis) for drinking nectar.

蝴蝶和蛾的区别

Difference between butterflies and moths

想区分蝴蝶和蛾可是有点难哟，因为它们都长着美丽的翅膀，会翩翩起舞。可是，虽然它们看起来很像，但还是有不少区别的。当你在花园里寻找蝴蝶时，记住这几个特点……

It can be hard to tell the difference between a butterfly and a moth because they both have beautiful wings to help them fly. While they may look alike, they have a few differences. Here are some things to look out for when searching for butterflies in your garden...

蛾的身体比较短，有很多毛。

Moths bodies are short and hairy.

蝴蝶的颜色比蛾更多彩、更漂亮。

Butterflies tend to be more colourful than moths.

蝴蝶

Butterfly

大多数蝴蝶夜晚休息，白天出来活动。

Most butterflies rest at night and come out during the day.

蝴蝶和蛾的翅膀都有细小的鳞片。蛾的翅膀通常更小，蝴蝶的翅膀更大。

Both have wings that are covered in tiny scales. Moths wings are usually shorter while butterflies have larger wings.

蝴蝶落下的时候，翅膀是并拢、竖起的。蛾落下的时候，翅膀是平放、展开的。

Butterflies rest with their wings closed and upright, whereas moths rest with their wings open and flat.

世界各地的蝴蝶

Butterflies around the world

世界上有 20 多万种美丽的蝴蝶和蛾。你见过哪些?

There are more than 200,000 species of spectacular butterflies and moths around the world. Which ones have you seen?

大黄带凤蝶
Giant swallowtail

斑马凤蝶
Zebra swallowtail

君主斑蝶 Monarch

东部虎纹凤蝶
Eastern tiger swallowtail

北美洲
North America

英国
UK

琉璃灰蝶
Holly blue

白钩蛱蝶
Comma

暗脉菜粉蝶
Green-veined white

南美洲
South America

橙斑黑蛱蝶
Grecian shoemaker

光明女神闪蝶
Morpho helena

在你家花园里，见过我的哪些小伙伴?

Can you see any of us in your garden?

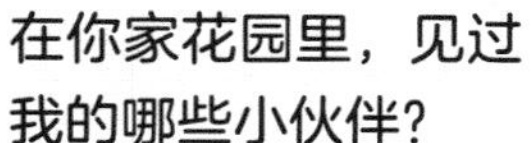

孔雀
蛱蝶
Peacock
红灰蝶
Small copper
红颈鸟翼凤蝶
Rajah Brooke's birdwing
紫闪蛱蝶
Purple
emperor
孔雀眼蛱蝶
Peacock
pansy
青凤蝶
Common
bluebottle
欧洲
Europe
亚洲
Asia
非洲
Africa
达摩凤蝶
Citrus swallowtail
小红蛱蝶
Painted lady
非洲绿带凤蝶
Green-banded
swallowtail
金斑蛱蝶
Danaid
eggfly
优红蛱蝶
Red admiral
歌利亚鸟翼凤蝶
Goliath birdwing
大洋洲
Oceania

鸣　谢

本出版社感谢以下机构提供照片使用权：

(a= 上方；b= 下方；c= 中间；f= 底图；l= 左侧；r= 右侧；t= 顶端)

32 Alamy Stock Photo: PhotoSpin,Inc (tr). **Dreamstime.com:** Jens Stolt / Jpsdk (cra, c). **33 Dorling Kindersley:** Natural History Museum, London (br). **Dreamstime.com:** Oleksandr Shpak (clb). **Getty Images / iStock:** Antagain (tc). **PunchStock:** Corbis (bl). **34 123RF.com:** Vassiliy Prikhodko (cb); Oksana Tkachuk (br). **PunchStock:** Westend61 (crb). **35 123RF.com:** Vassiliy Prikhodko (crb); Oksana Tkachuk (cr); Richard E Leighton Jr (cla). **Alamy Stock Photo:** PhotoSpin,Inc (tl). **Dorling Kindersley:** Natural History Museum, London (clb/butterfly, cra). **Dreamstime.com:** Stephan Bock / Tunedin61 (clb). **36 Dorling Kindersley:** Natural History Museum, London (r). **37 Dorling Kindersley:** Natural History Museum, London (l). **Dreamstime.com:** Gorodok495 (bc); Matee Nuserm (br). **38 Dorling Kindersley:** Natural History Museum, London (cla, clb, cra, ca, crb). **Dreamstime.com:** Eivaisla (br); Stephanie Frey (cl). **PunchStock:** Corbis (c). **39 123RF.com:** Richard E Leighton Jr (cr). **Dorling Kindersley:** Natural History Museum, London (tl, tr, ftr, fcra, cb). **Dreamstime.com:** Alslutsky (bl); Feathercollector (cra, crb/ Goliath); Matee Nuserm (clb); Domiciano Pablo Romero Franco (bc). **Getty Images / iStock:** epantha (crb). **PunchStock:** Westend61 (tc).

其余图片版权归英国 DK 公司所有，更多信息请访问 www.dkimages.com。

关于作者和绘者

本和弗朗西斯是一对夫妻，住在英国泰恩河畔纽卡斯尔。他们倾心于帮助家中花园里的野生动物。仲夏的夜晚，弗朗西斯醒来，有了一个灵感——创作鼓励小朋友加入他们的行动的系列童书。

弗朗西斯创作故事，本画插图他们引领小读者进入一个栩栩如生的野生动物世界，快来欢迎花园里的小客人：小刺猬罗里、小麻雀罗芮、小蜜蜂罗丝、小蝴蝶罗克西。